# I'm a Triceratops

my dinosaur adventure

Cherry Lake Press

Published in the United States of America by Cherry Lake Publishing
Ann Arbor, Michigan
www.cherrylakepublishing.com

Reading Adviser: Marla Conn, MS, Ed., Literacy specialist, Read-Ability, Inc.
Content Adviser: Kierstin Rosenbach, Ph.D. Candidate, Vertebrate Paleontology, University of Michigan
Book Designer: Jennifer Wahl
Illustrator: Jeff Bane

Photo Credits: © David Roland/Shutterstock.com, 5; © Daniel Andis/Shutterstock.com, 7; © CSA-Printstock/iStock.com, 9; © JONATHAN PLEDGER/Shutterstock.com, 11; © leonello/iStock.com, 13; © dugdax/Shutterstock.com, 15; © ugurhan/iStock.com, 17, 23; © Herschel Hoffmeyer/Shutterstock.com, 19; © Elenarts/Shutterstock.com, 21; Cover, 2-3, 10, 12, 22, 24, Jeff Bane



Library of Congress Cataloging-in-Publication Data

Names: Nelson, Jake, author. | Bane, Jeff, 1957- illustrator.
Title: I'm a triceratops / Jake Nelson; illustrator, Jeff Bane.
Description: Ann Arbor, Michigan: Cherry Lake Publishing, [2021] | Series: My dinosaur adventure | Includes index. | Audience: Grades K-1
Identifiers: LCCN 2020002473 (print) | LCCN 2020002474 (ebook) | ISBN 9781534168480 (hardcover) | ISBN 9781534170162 (paperback) | ISBN 9781534172005 (pdf) | ISBN 9781534173842 (ebook)
Subjects: LCSH: Triceratops--Juvenile literature.
Classification: LCC QE862.O65 N454 2021 (print) | LCC QE862.O65 (ebook) | DDC 567.915/8--dc23
LC record available at https://lccn.loc.gov/2020002473
LC ebook record available at https://lccn.loc.gov/2020002474

Printed in the United States of America
Corporate Graphics

# table of contents



**About the author:** Jake Nelson was born and raised in Minnesota, where he enjoys everything from watching the Twins at Target Field to strolling along the shore of Lake Superior. He writes books, blogs, and content for the web.

**About the illustrator:** Jeff Bane and his two business partners own a studio along the American River in Folsom, California, home of the 1849 Gold Rush. When Jeff's not sketching or illustrating for clients, he's either swimming or kayaking in the river to relax.

Oh, hello there!
I am a Triceratops.

I am 30 feet (9 meters) long.

I weigh as much as 11 horses!

I lived on Earth during the **Mesozoic era**.

It was about 66 million years ago.

I have three horns.

One is on my nose. Two are on my head.

My name means "three-horned face."

Tri means "three." A tricycle has three tires.

I like to eat plants, like ferns and leaves.

That makes me an **herbivore**.

Other dinosaurs are called **carnivores**.

They hunt **prey**.

That is why I need my horns.

They help me defend myself from **predators**.

**Paleontologists** also think I use my horns to knock down trees.

That way, I can eat the leaves.

My horns are very useful.

They are what makes me a Triceratops!

glossary & index

## glossary

**carnivores** (KAHR-nuh-vorz) creatures that only eat other living things, like animals or bugs

**herbivore** (HUR-buh-vor) a creature that only eats plants, like grass and leaves

**Mesozoic era** (mez-uh-ZOH-ik ER-uh) the period of time when dinosaurs lived on Earth, between 245 million and 66 million years ago

**paleontologists** (pay-lee-uhn-TAH-luh-jists) scientists who study dinosaurs

**predators** (PRED-uh-turz) creatures that hunt and eat another creature

**prey** (PRAY) a creature that is hunted by another creature

## index